A DOCTOR'S GUIDE TO HELPING YOURSELF WITH BIOCHEMIC TISSUE SALTS

A doctor with thirty year's experience of practising biochemic tissue salt medicine explains all about tissue salt therapy and provides a comprehensive therapeutic index with advice on self-help.

A Doctor's Guide to Helping Yourself with BIOCHEMIC TISSUE SALTS

by

Peter Gilbert

B.A.,M.B.,B.S.,M.R.C.S.,L.R.C.P.

THORSONS PUBLISHERS LIMITED
Wellingborough, Northamptonshire

First published 1984

British Library Cataloguing in Publication Data

Gilbert, Peter
A Doctor's guide to helping yourself with biochemic tissue salts.
1. Health
I. Title
613 RA776.6

ISBN 0-7225-0925-1

Printed in Great Britain by
Richard Clay (The Chaucer Press) Ltd,
Bungay, Suffolk

CONTENTS

Although a great deal can be done by self-help alone, medical help should always be sensibly sought as necessary.

DEDICATION

To my late father, Dr Henry Gilbert, the pioneer in this field in the U.K. to whom I will always be indebted, and to New Era Laboratories Limited who have given me every assistance in the preparation of this book.

INTRODUCTION

The medical profession, for the most part, can only supply us with drugs and operations for our ailments. These of course have their vital place, and the caring medical profession is essential to the well-being of our people, but many of us doctors are becoming increasingly worried about drug side-effects, and feel that for many people something less drastic would be fully adequate to help so many ailments. Thus there has come about a revival of alternative medicine; medicine that is complementary to and has no quarrel with allopathic medicine but can be used frequently as an alternative or often with benefit as an additional therapy. Of all the methods of such self-help the safest in my opinion is the biochemic tissue salts, and this book is about just these; that is the twelve biochemic tissue salts and the various combination remedies derived from them.

As people become more health conscious they are also becoming increasingly aware that they can themselves influence their health by various positive means which are not covered by western technological medicine. It is because allopathic medicine only supplies drugs, and because prevention of ailments is not even a starter with such an approach, that people have looked around for help elsewhere. This can be full of pitfalls, and it is nice to know that the tissue salts are so totally safe that in their case there can be no grounds for anxiety.

This search for self-help has of course given rise to the health food movement, with the trend towards natural wholefood products free of additives. Such a movement was not so long ago thought by many to be cranky and faddy, but now many doctors realize the great value of wholefoods. However, most families still do nothing by way of self-help for the minor disorders they meet in everyday life. My contention is that they should look to biochemic tissue salts as a means of safe self-help. These are now freely available in Health Food Stores and elsewhere.

I feel that of all the systems of self-help available none can compete with biochemic tissue salts for their combination of simplicity and effectiveness — and it is being so safe and effective which makes them the most admirable system for effective home treatment. I have used biochemic tissue salts extensively for over thirty years in my medical practice and can personally vouch for their efficacy. People, I feel, should help themselves, that is with all due safeguard take a hand in getting well and keeping well. That is the purpose of this book.

1.

THE NEED FOR TISSUE SALTS

Most doctors, I am sure, will agree with me that food is really the basis of our health. Where does all our food come from? A great deal of it comes from the soil, and so our health is linked with the health of the soil. The substances in the earth which we use and are of most interest to us here are the minerals. One could say that a healthy mineral balance in the soil from which our food comes is the basis of health. So we see that a fertile mineral-rich balanced soil is the necessary foundation for good health.

In this country soil fertility from the mineral point of view has deteriorated so that the soil does not yield sufficiently healthy and sufficiently health-giving foods by which alone real health is obtainable. This poor health is not as obvious as it should be, however, because medical science has produced a great deal of what may be called artificial or imitation health. Today very many men and women have to take drugs continually to help them lead an apparently normal life, and we all know that the continual use of drugs is, to put it mildly, dangerous.

Powerful drugs are to be avoided unless there is no alternative. But even if there is no alternative such people can benefit from the biochemic tissue salts taken over a long period of time. Then with gradually improving health they can hopefully, eventually, be weaned off the otherwise essential drugs. It is as well to remember that a person who appears to be in good health may be in a condition of latent ill health because of mineral deficiencies. Even with a perfectly balanced diet and abundant use of health foods we may not have sufficient minerals in us, and with such a deficiency we begin to lack the power to absorb nourishment properly from food. A truly vicious circle. However, tissue salts can be absorbed and can be the passport to recovery. This is because the fine minerals of the tissue salts enable the rougher minerals of our present day food to follow through, so in time correcting deficiencies and enabling a return to robust health. That is to say that long-term treatment is often required with the tissue salts, and is indeed most beneficial.

So we see that minerals are necessary for health all along the line, and these minerals constitute essential links for life wherever living processes occur. They may be small in amount in the body but they play a disproportionately large part in the routine chemical processes of all living things. Some of these minerals, which are required in such minute amounts as a few parts per million, are vital to the well-being of the biological systems of living organisms. Life could not exist without the presence of mineral elements in our bodies. If there is any deficiency of the integrating mineral elements then there will be ill health and such ill health can of course be corrected by giving the necessary missing minerals if — and I repeat if — these missing minerals are properly prepared. This matter of the vital importance of proper preparation of the biochemic tissue salts will be covered in the next chapter.

Now, what tissue salt to choose in a given case? This may seem somewhat confusing with twelve tissue salts and eighteen combinations of these tissue salts, plus other special products, such as *Elasto, Zief* and *Nervone*, but with a little time it is possible for the interested person to decide the few from the list that could apply to his or her case, and then to look again at those few to decide which one has the greatest indications for him or her. In this way the most appropriate product can be arrived at. Regular users really do get skilful and the benefits certainly more than justify the time taken in choosing. The therapeutic index that appears later in this book will indicate what to use for given conditions. Such self-help was frowned upon by the medical profession in the past but now, with all due safeguards, such self-help is being encouraged as one way of getting the vast costs of the National Health Service reduced.

Many people find these tissue salts the perfect answer to keeping well, that is they use them prophylactically, prevention being better than cure.

Every nutritionist admits that an individual's requirement of minerals varies widely from one to another, and it has been found now that they vary even more than was at first thought. Not only do requirements vary from person to person but also, from time to time, the mineral requirements change in individuals. For example, say you were to have a very trying and worrying day. In such circumstances you would use up more phosphates than you would need on a calm day. In such a case *New Era Kali.Phos. 6X* would be helpful or *Nervone* if you are really down.

Another matter I feel worth remembering is that the mineral elements in the body are a deciding factor in relation to resistance to bacterial and viral diseases. The minerals control for example the defence enzymes. These enzymes cannot work on their own. They need the presence of substances called activators. The main activators involve the mineral tissue salts, so the use of the tissue salts can help to prevent ill health as well as playing their part in restoring health. Because mineral lack over some years can eventually give rise to ill health it is hardly surprising, therefore, that for tissue salts to be effective they have to be taken for a correspondingly lengthy period of time; that is, you may not have realized that tissue salts can be a long-term help as well as being suitable for short term therapy for more acute conditions.

People I feel should be made aware that they must not take for granted that a good mixed diet will be all they need for good health. First, what they need must be taken into the body but this is no guarantee that such healthy foods will be either absorbed or subsequently used properly. The tissue salts help to ensure better absorption of nutrients, and then to help such nutrients to be used properly in the body for maximum health to be achieved. And further to help ensure that waste products are properly eliminated and do not harmfully accrue in the tissues. The latter, that is accumulation of waste products in our tissues, is thought by many to be one of the reasons for the whole rheumaticky range of ailments. So we see that on numerous counts there is the need for tissue salts.

2.

HOW THE NEW ERA BIOCHEMIC TISSUE SALTS REMEDIES ARE MADE

The tissue salt remedies are made by using a process of trituration and one part of the chosen mineral is ground very finely indeed over many hours in a trituration grinding machine with nine parts of sugar of milk (lactose). This brings the salt from zero to lx potency. Then one part of this is similarly ground by the trituration machine with a further nine parts of lactose to produce 2x potency and this is repeated again and again until 6x strength is reached.

It is the 6x potency which is used in all the *New Era* biochemic tissue salts, and the biochemic combination of these, as at this level of 1 part of element in a total matrix-million the trace element effect has been found to be the most effective therapeutically and totally safe for self-help therapy. But why grind these tissue salts so finely with this trituration process we speak about? The answer lies with the geological origin of soil. Rocks over many millions of years are eroded by nature to produce soils and its erosion over such a large period of time produces a very fine particle size of mineral indeed. These dissolved fine particles of mineral played their part in the origin of life itself, and the need for such fine particles of minerals remains with us to this day as our evolutionary heritage.

It is not a matter of argument, it is a basic fact that our body cells need minerals in very fine particle size. The fine trituration of the tissue salts mimics what nature took so many thousands of years to achieve. With overcropping of soils these very fine particle-sized minerals become less in amount and we are left with larger mineral particles which we find more difficult to manage in life processes. When we do manage we are not at our optimum, and if there is real difficulty then we become unwell and ailments follow. The fine particle minerals, when present, enable us to make better use of the rougher minerals. So you see that it follows that the simple tissue salts can be vital for our welfare but are ineffective unless of very fine particle size and easily absorbable into our systems. To ensure this they must be made in such a way as to fulfil these requirements.

My assessment is that there is no better place known to me for the manufacture of these vital tissue salts than New Era Laboratories Limited in London. It is nice to know that you can have total confidence in these products which are made with great care. Subsequent to the lengthy trituration process these powders are turned into tablet form, and tableting of the salts can make or mar the product. If solid tablets are made, or if ready bought compressed tablets have added to them merely a tincture then the product is ineffective from the biochemic tissue salt point of view.

It is important to go to all the trouble to make the tablet a soft pellet. It then dissolves in the mouth, that is through the surface of the tongue and adjacent mucosa and so is absorbed directly into the bloodstream. If the tablets have to be chewed or swallowed then they are altered by the digestive tract which may negate the good that would otherwise be done. So the type of tablet made is essential for the product to work properly. Meticulous attention to details is paid at New Era Laboratories to ensure the correct type of easily assimilable tablets and they are made under strict Department of Health pharmaceutical preparation regulations.

So you see that great importance attaches to the trituration of the minerals for the tissue salts to be really effective. Then, when the right substance is offered to the body in a fine particle size and in an easily absorbable tablet the process of improving health can move steadily forward.

3.
NEW ERA REMEDIES AVAILABLE

The *New Era* remedies available are all in the 6x potency, as mentioned in the previous section, and are available as the twelve *New Era* biochemic tissue salts, the eighteen combinations and also other preparations which have stood the test of time such as *Elasto, Nervone* and *Zief.* In addition to these there are at present four vitamin preparations containing tissue salts, and a range of skin preparations also containing vital biochemic tissue salts. The vitamin and beauty preparations will be considered in later sections.

Also, towards the end of the book, is the therapeutic index which indicates how these products can be used effectively to assist in so many ailments.

But first let us consider these remedies in turn and their general use and the usual nomenclature which has been established over the years to easily recognize each one.

To begin with, the twelve biochemic tissue salts at 6x potency are known by numbers from 1-12. The order being their Latin names in alphabetical order. It is this Latin name which historically has been used to indicate that the product has been finely ground, to differentiate them from the usual chemical name of the minerals when they have not been ground finely by the trituration process. For example, No. 12 is the mineral silicon dioxide, but when triturated it is called *silica* and the 6x designation indicates the level of such trituration.

In order the twelve tissue salts are:

1. *(Calc. Fluor.)*
2. *(Calc. Phos.)*
3. (*Calc. Sulph.*)
4. (*Ferr. Phos.*)
5. (*Kali. Mur.*)
6. (*Kali. Phos.*)
7. (*Kali. Sulph.*)
8. (*Mag. Phos.*)
9. (*Nat. Mur.*)
10. (*Nat. Phos.*)
11. (*Nat. Sulph.*)
12. (*Silica*)

All being in 6x potency.

Now we will consider each of the twelve biochemic tissue salts in turn taking them in the order of the previous list given, and of course a supply of all twelve of these should always be readily available in the home.

No. 1 (*Calc. Fluor.* 6x)
In nature this occurs, as do all the other minerals, in the earth and in the rocks. In the body this mineral occurs in:

bones
teeth
walls of blood vessels and all connective 'holding' tissues

Thus deficiency of this salt leads to upset of the above body components. Remember, that these can of course lead in turn to other upsets if neglected. Thus more than one salt may be needed to be given in a particular case, but *Calc. Fluor.* is the one to use for maintaining tissue elasticity. The chief indications for *Calc. Fluor.* are as follows:

varicose veins
varicose ulcers
piles
over relaxed tissues
flabbiness or giving way and sagging of tissues, e.g. hernias and prolapse
poor teeth
late development of teeth in infants and children
deficient enamel

No. 2 (*Calc. Phos.* 6x)
This mineral occurs in bones and teeth and also in soft tissues but is mainly concerned with bony structures generally. Deficiency of this salt is seen in the following conditions:

slow healing of fractures

bony deformities
delay in teething and general teething problems
some types of anaemia (calcium is important for the proper formation of blood)
poor nutrition and digestion
coldness
cramps
chilblains
liability to colds and catarrh
children outgrowing their strength
chronic tonsillitis
some skin diseases (catarrhal type)
polypus
coccydynia
hypochondriasis (with *Kali. Phos.* No. 6)

In several of the above, including the latter, other remedies should often also be used. This will be covered later when I deal with the *New Era* combination remedies.

With *Calc. Phos.* deficiency state, subjects are usually made worse by coldness, coffee, tobacco, and excessive self-contemplation of symptoms. This latter factor of hypochondriasis is due to the fact that the phosphorus balance is essential for the proper functioning of the nervous system. See No. 6 *Kali. Phos.* 6x later and the remedy *Nervone.*

Calc. Phos. subjects are usually helped by warm dry sunny weather and by rest — preferably bed rest.

No. 3 (*Calc. Sulph.* 6x)

In the body this mineral occurs in connective tissue, as a blood constituent and also in the liver cells. The function in the latter case is the removal of worn out blood cells from the circulating bloodstream. A deficiency of *Calc. Sulph.* impairs this cleaning activity and related problems arise as well as disorders of connective tissue. With a *Calc. Sulph.* deficiency the following conditions are seen:

pimples during adolescence
boils
skin eruptions
skin slow to heal
catarrh
dandruff
falling hair
vertigo with nausea
sore lips
gum boils
neuralgia
frontal headaches (particularly so in elderly people)
pancreatic upsets
liver upsets (with No. 11 *Nat. Sulph.* 6x)
kidney upsets

No. 4 (*Ferr. Phos.* 6x)
This salt is found in all the tissues of the body, but mainly in the red blood cells. The iron is concerned with the oxygen-carrying capacity of the blood. Its use in muscular coats of blood vessels becomes apparent when the effects of deficiency are seen — this leads to relaxation of the walls of blood vessels which in turn leads to congestion and inflammation. Thus *Ferr. Phos.* is used for all inflammations, that is

all ailments ending with 'itis', e.g. bronchitis. Thus *Ferr. Phos.* 6x will be helpful in the first stages of these ailments and if the condition is beyond this stage then *Kali. Mur.* 6x should be used as well (see later).

Ferr. Phos. is a great children's remedy. It is helpful in healing their so often worrying conditions and can frequently help to put these right before the condition has advanced to the stage of diagnosis in the orthodox sense. But, as with all conditions, always see your doctor if there is the slightest doubt.

To summarize, *Ferr. Phos.* 6x is used in the following conditions:

all the minor respiratory disorders
childhood illnesses (e.g. measles, scarlet fever, etc.)
first stage of all inflammations and fevers
congestions
haemorrhage
nose bleed
excessive periods
throbbing, congested headache
inflammatory rheumatism
coughs and colds
chills, 'feverishness'
chestiness (in alternation with *Kali. Mur.* 6x)

This remedy regulates the bowel — that is from either looseness or from constipation — and is most useful in children in this respect.

Ferr. Phos. is the most frequently used of the tissue salts traditionally, but now closely followed by *Kali. Phos.* 6x (see later) in these stressful times.

A little powdered *Ferr. Phos.* 6x (they are easily crushed to a powder) when applied externally can help staunch blood flow from small cuts and promote clean and rapid healing.

No. 5 (*Kali. Mur.* 6x)
A deficiency of this salt affects the fibrin in the body, that is the cells below the skin that exude thus producing the picture seen in the second stage of inflammation. *Kali. Mur.* is thus used for the second stage of all inflammatory illnesses. The salt is used for the following conditions:

- second stage of inflammation of all 'itis' illnesses
- minor respiratory disorders
- coughs, cold symptoms, wheeziness
- chestiness (in alternation with *Ferr. Phos.* 6x for children's feverish colds)
- prior to vaccination or immunization to help eliminate undesirable side-effects
- chickenpox
- scarlet fever
- mumps
- measles
- white/grey coating of tongue
- catarrh
- eczema — especially infantile
- warts
- acne
- constipation (in liverish states and in pregnancy)
- diarrhoea due to fatty foods
- piles

menorrhagia (but always see your doctor)
leucorrhoea
shingles
burns
scalds

With a *Kali. Mur.* 6x deficiency state, patients usually find they are worse with fatty foods.

No. 6 (*Kali. Phos.* 6x)

Kali. Phos. 6x is the remedy for the nervous system and is the great nerve soother — phosphorus being the chief nerve mineral. It is used for all nervous disorders and those that are so often called 'neurotic' illnesses, and treating them along these lines is most helpful. From the presence of *Kali. Phos.* in other tissues of the body comes its other uses. The salt is used in the following conditions:

all temporary nerviness
melancholia
hysteria
highly strung
fearfulness
despair
timidity and shyness
loss of mental and nerve power
nervous debility
emotional strain
excessive blushing
neuritis
incontinence or retention of urine from *nervous* causes
nightmares — all 'unjustified' fears, e.g. of water.

inability to sleep
nervous indigestion
nervous diarrhoea
nervous headache
alopecia
nervous asthma
menstrual colic — spasms, cramps etc.
ineffectual labour pains from anxiety
sexual incompetence and frigidity

Patients requiring this remedy are usually made worse by noises and moving about, and when left alone too long, and also by over-excitement. They are helped by soothing company and conversation on uncontroversial topics.

The remedy is useful at some stage in most illnesses, e.g. for the lack of sleep and panic so often seen; indeed it can be very useful for the relatives of patients who may very well need the help of a salt almost as much as the patient.

No. 7 (*Kali. Sulph.* 6x)

This tissue salt is found in the external layers of epithelial membrane, such as the skin, and maintains skin condition. Disturbance of this causes yellowish catarrh and shedding of surface epithelia. It is thus concerned with the third stage of inflammation of all inflammatory illnesses — the 'itis' diseases. It is also concerned with oxygen carriage in the body. *Kali. Sulph.* is used for the following conditions:

third stage of all inflammations
bronchitis
yellow coating of tongue

thick yellow mucous catarrh
whooping cough
gastric catarrh
intestinal catarrh
asthma
colic (if response to *Mag. Phos.* is poor)
menstrual disorders
to help maintain healthy hair
dandruff
foul breath
measles
eczema
psoriasis
brittle nails
minor skin eruptions with scaling or sticky exudation
flashes of heat and chilliness
giddiness of inflammatory type
palpitations
headache

Patients requiring *Kali. Sulph.* are frequently worse in hot weather.

No. 8 (*Mag. Phos.* 6x)

Mag. Phos. is found in similar tissues to *Calc. Phos.* 6x, i.e. bones, teeth and nerve tissues, as well as in blood vessels and muscles, but basically it is a soft tissue salt. Deficiency of this salt leads to cramp-like conditions and collicky states and so it is known as the anti-spasmodic tissue salt.

Mag. Phos. is used for the following conditions:

muscle cramps and spasms
minor occasional pains
hiccups
spasmodic shivers and twitching
intermittent retention of urine and bladder spasm
enlarged prostate
writer's cramp and similar conditions
stuttering
crampy labour pains
painful menstruation
ovarian neuralgia
gallstone colic
kidney stone colic
teething in infants
constipation in infants
flatulence
intercostal neuralgia
headaches
toothache when pains are sharp, shooting, boring
rheumatic pains
neuralgia generally

Mag. Phos. subjects are often lean, nervous, people. Pains are usually worse from cold and touch and helped by warmth, pressure and bending.

No. 9 (*Nat. Mur.* 6x)

Nat. Mur. is based on a mineral called sodium chloride, or common table salt. We need one gram of the crude salt per day in Britain but the average intake is 10 grams daily. This excessive intake creates havoc with many people, and the fine particle minerals of tissue salts remedy No. 9, that is *Nat. Mur.* 6x, helps to put right the sodium imbalance in the body.

Nat. Mur. occurs in all tissue fluids in the body and is often regarded as the most important of the twelve salts. It is a distributor and controller of water throughout the body. *Nat. Mur.* is used in the following conditions:

circulation problems
shock
watery vomiting
diarrhoea
minor haemorrhage
anaemia (always see your doctor)
watery colds with flow of tears and runny nose
loss of smell
loss of taste
prolapse (with *Calc. Fluor.*), e.g. shingles, herpes and blisters
insect stings and bites generally
thin watery milk in lactation
excessive size of breasts in pregnancy
hydrocele (in the scrotum)
excessive tears
excessive salivation
teething with excessive salivation
water brash
sneezing
hay fever
influenza
asthma

constipation (dry stools)
headaches (early morning type)
hysteria
sterility
nettle rash
chronic eczema
acne
greasy skin
ulcer of gums
gout
sciatica
sunstroke

No. 10 (*Nat. Phos.* 6x)
This salt occurs in the inter-cellular fluids and in the body tissues generally. It has two main types of action in the body. It controls acid generally, sometimes called the 'acid neutralizer', and it helps deal with fatty acids.

Its first action has its uses in dealing with uric acid and lactic acid etc., thus helping to prevent or treat acid states.

The second main function is seen in the help it can give in dyspepsia due to excessive intake or improper usage of fat.

Nat. Phos. 6x is used for the following conditions:
all acid states of the blood stream 'uric acid dialysis'
rheumatism of joints
rheumatic arthritis
gout
acid taste

to help prevention of gallstones etc.
sick headaches
giddiness
conjunctivitis
itching of nose
red, blotchy face
sea sickness (with *Kali. Phos.)*
sour breath
grinding of teeth during sleep
yellow coated tongue
catarrh and thick yellow mucous
heartburn
nausea
morning sickness of pregnancy
gastric indigestion
sour flatulence
loss of appetite
constipation
diarrhoea
acidity
incontinence from acidity
sterility from acidity
leucorrhoea — sour smelling
sleeplessness from itching in acid states
rheumatic pain tendency

Subjects requiring *Nat. Phos.* are made worse by fats and sugars (sweets in children). Reducing the intake of these is usually found helpful.

No. 11 (*Nat. Sulph.* 6x)

This salt occurs mainly in the intercellular tissues, and it has an affect on the water content of the body, but in a rather different way from *Nat. Mur. Nat. Sulph.* appears to withdraw water and so it helps the elimination of effete cells. *Nat. Sulph.* is used mainly in the following conditions:

to help with the body water balance
billiousness of watery nature
vomiting in pregnancy
diarrhoea
constipation
tongue grey or greenish/brown
bitter taste
liver upsets
gall bladder upsets
kidney upsets
pancreatic upsets
rheumatism in watery subjects
gout in watery subjects
asthma of a watery nature
bronchial catarrh
flu symptoms
hay fever
warts
ear noises and earache from fluid retention
flatulence and colic
distended stomach
queasiness
digestive upsets
hydrocele (in the scrotum)

Subjects needing *Nat. Sulph* are usually worse in damp weather, and feel better in dry conditions, hot or cold.

No. 12 (*Silica* 6x)

This salt occurs in the connective tissue, and disturbance of its balance affects the nervous system from its presence in the coverings of nerve fibres. Disturbance of the silica content of the body usually ends in pus formation and *Silica* 6x helps to promote and push out such pus (note that *Calc. Sulph.* has a rather different action). Therefore *Silica* 6x can be thought of as a conditioner and a cleanser. It is also a preventative in premature ageing — a lack of this salt causing premature ageing by atrophy of tissues.

In our food, silica is found in wholemeal bread and is the 'grit' of our diet.

Silica 6x is used in the following conditions:

- lack of 'grit' (i.e. stamina both physical and mental)
- absent-mindedness
- poor memory (see *Kali. Phos.*)
- foot sweats
- sweats generally if of an unpleasant nature
- alcoholic intolerance
- falling out of hair
- brittle nails (in alternation with *Kali. Sulph.*)
- eye strain
- asthma from dust
- pimples and spots
- styes
- fissures
- boils
- chronic bronchitis
- coccydynia
- alopecia

whitlows
ingrowing toe-nails
premature ageing for the elderly

Patients needing silica are usually worse at night and rom feet becoming cold. They are usually helped by warm atmosphere and from hot baths.

* * *

This concludes the list of the twelve tissue salts. To ake them follow the instructions on the container or ollow the advice of a biochemic physician. You annot take an overdose — more than the body needs would just be a waste and nothing more, so there is total safety.

4.

THE NEW ERA COMBINATION TISSUE REMEDIES

For many ailments more than one tissue salt is indicated as you will have seen from the previous section on the twelve salts. So, to simplify treatment, the combination remedies have been designed to cover in one tablet what is needed for frequently found conditions. These remedies are the result of many years of medical experience, and a comprehensive range is made by New Era Laboratories Limited.

The range is given below with some indication of their use, but for a more extensive coverage see the Therapeutic Index later in this book.

Combination Remedy A

This remedy contains the tissue salts *Ferr. Phos.* 6x, *Kali. Phos.* 6x, and *Mag. Phos.* 6x. It is used for: sciatica, neuralgia and neuritis.

ombination Remedy B

his remedy contains the tissue salts *Calc. Phos.* 6x, *ali. Phos.* 6x and *Ferr. Phos.* 6x. It is used for: eneral debility, edginess, nervous exhaustion and onvalescence.

ombination Remedy C

his remedy contains the tissue salts *Mag. Phos.* 6x, *at. Phos.* 6x, *Nat. Sulph.* 6x and *Silica* 6x. It is used r: acidity, heartburn, and dyspepsia.

ombination Remedy D

his contains: *Kali. Mur.* 6x, *Kali. Sulph.* 6x, *Calc. ulph.* 6x, *Silica* 6x. It is used for: minor skin ilments, scalp eruptions, eczema, acne and scaling f the skin.

ombination Remedy E

his contains: *Calc. Phos.* 6x, *Mag. Phos.* 6x, *Nat. hos.* 6x, *Nat. Sulph.* 6x. It is used for: flatulence, olicky pains and indigestion.

ombination Remedy F

his contains: *Kali. Phos.* 6x, *Nat. Mur.* 6x, *Silica* 6x. is used for: nervous headaches and migraine.

ombination Remedy G

his contains: *Calc. Fluor.* 6x, *Calc. Phos.* 6x, *Kali. hos.* 6x, *Nat. Mur.* 6x. It is used for: backache, mbago, piles.

ombination Remedy H

his contains: *Mag. Phos.* 6x, *Nat. Mur.* 6x, *Silica* 6x.

It is used for: hay fever and allied conditions.

Combination Remedy I

This contains: *Ferr. Phos.* 6x, *Kali. Sulph.* 6x, *M*
Phos. 6x. It is used for: fibrositis and muscular pai

Combination Remedy J

This contains: *Ferr. Phos.* 6x, *Kali. Mur.* 6x, *N*
Mur. 6x. It is used for: coughs, colds, and chestin
and is the autumn and winter seasonal remedy.

Combination Remedy K

This contains: *Kali. Sulph.* 6x, *Nat. Mur.* 6x, *Silica* 6
It is used for: brittle nails and falling hair.

Combination Remedy L

This contains: *Calc. Fluor.* 6x, *Ferr. Phos.* 6x, *N*
Mur. 6x. It is used for: sedentary life-style, toning t
tissues and is a natural tonic.

Combination Remedy M

This contains: *Calc. Phos.* 6x, *Kali. Mur.* 6x, *N*
Phos. 6x, *Nat. Sulph.* 6x. It is used for: rheuma
pains.

Combination Remedy N

This contains: *Calc. Phos.* 6x, *Kali. Mur.* 6x, *Ka*
Phos. 6x, *Mag. Phos.* 6x. It is used for: menstru
pain.

Combination Remedy P

This contains: *Calc. Fluor.* 6x, *Calc. Phos.* 6x, *Ka*

hos. 6x, *Mag. Phos.* 6x. It is used for: aching feet and gs.

ombination Remedy Q

'his contains: *Ferr. Phos.* 6x, *Kali. Mur.* 6x, *Kali ulph.* 6x, *Nat. Mur.* 6x. It is used for: catarrh and nus disorders.

ombination Remedy R

'his contains: *Calc. Fluor.* 6x, *Calc. Phos.* 6x, *Ferr. hos.* 6x, *Mag. Phos.* 6x, *Silica* 6x. It is used for: nfant's teething pains.

ombination Remedy S

'his contains: *Kali. Mur.* 6x, *Nat. Phos.* 6x, *Nat. ulph.* 6x. It is used for: stomach upsets, biliousness, ueasiness, sick headaches. This is an excellent ummer seasonal remedy.

urther Special Combinations

'urther special combinations available that have roved their worth over many years are: *Elasto, Vervone* and *Zief.* For their uses see the Therapeutic ndex but some comments may be made here.

lasto

This contains: *Calc. Fluor.* 6x, *Calc. Phos.* 6x, *Ferr. hos.* 6x, *Mag. Phos.* 6x. It is used for: varicose veins nd aching legs.

Vervone

This contains: *Calc. Phos.* 6x, *Kali. Mur.* 6x, *Kali.*

Phos. 6x, *Mag. Phos.* 6x and *Nat. Phos.* 6x. It is us
for: nervous disability and nerve pains.

Zief
This contains: *Ferr. Phos.* 6x, *Nat. Phos.* 6x, *N*
Sulph. 6x, *Silica* 6x. It is used for: rheumatic pain

Note: Elasto derives its name from its beneficial effe
on the elastic tissues of the body. *Nervone* is an excelle
safe and reliable remedy for a whole range of 'ner
troubles' and allied ailments. *Zief* is a combination
four tissue salts which are indicated for treatment
rheumatic conditions.

5.

THE NEW ERA VITAMIN RANGE

As with minerals, vitamins are important nutrients necessary for vital functions to proceed normally in the body, and a unique range of vitamin products is also made by *New Era* with the same care and attention as with their range of tissue salts, and indeed containing selected tissue salts.

Multi-vitamins
This in an all purpose multi-vitamin supplement providing the important standard vitamins and minerals in easy to swallow capsule form combined with *Ferr. Phos.* 6x, *Kali. Phos.* 6x, and *Mag. Phos.* 6x. The formula is:

Di-calcium phospate	200mg	Vitamin C	35mg
Vitamin E	35mg	Vitamin E	50mg
Iron	7mg	Nicotinamide	15mg
D calcium pantathenate	5mg	Choline Bi-tartrate	5mg

Riboflavin (vitamin B_2)	2mg	Thiamine (Vitamin B1)	3m
Pyridoxin (vitamin B_6)	0.5mg	Inositol	1m
Potassium sulphate	1mg	Magnesium sulphate	0.5m
Molybdenum	0.05mg	Copper	0.2m
Manganese	5mcg	Zinc	0.05m
Biotin	0.3mcg	Iodine	3mc
Potassium phosphate	0.12mcg	Iron phosphate	0.12mc
Vitamin A	1,250 i.u.	Magnesium phosphate	0.12mc
Vitamin B_{12}	1mcg		

Uses for these preparations are indicated in th Therapeutic Index later in the book, and also in thi chapter some idea is given of the uses of the individua items in this formulae.

Vitamin B Complex

This is the formula:

Vitamin B_1	2.0mg	Vitamin B_2	1.5m
Vitamin B_6	1.0mg	Vitamin B_{12}	1.0mc
Calcium pantothenate	4.0mg	Choline bitartrate	10.0m
Inositol	5.0mg	Nicotinamide	10m
Biotin	0.1mcg	Yeast	50.0m
Calcium phosphate	0.6mcg	Iron phosphate	0.06mc
Magnesium phosphate	0.06mcg	Potassium phosphate	0.06mc

Vitamin C Capsules

These have the formula:

Vitamin C	200mg	Calcium phosphate	0.09mc
Iron phosphate	0.09mcg	Magnesium phosphate	0.09mc
Potassium phosphate	0.09mcg		

Vitamin E Capsules

These have the formula:

Vitamin E	110mg	Iron phospate	0.09mc

Calcium phosphate 0.09mcg
Potassium phosphate 0.09mcg
Magnesium phosphate 0.09mcg

The need for supplementary vitamins has greatly increased over the last few years with present-day 'convenience' food lacking in essential nutrients for our full well-being. The following gives some indication of their uses and our need for them:

Vitamin A

This is essential for proper growth and development of bones and teeth, for vision, and for healthy skin. It protects the tissues of the respiratory tract and so helps combat infections. It is also necessary for the healthy function of the liver and reproductive organs.

Thiamine (Vitamin B_1)

Thiamine is essential for the health of the digestive tract and for a normal healthy nervous system and mentality.

Riboflavin (Vitamin B_2)

This is essential to proper enzyme formation and for normal growth and tissue formation, and for the efficient metabolism of carbohydrates, fats and protein. It is helpful in conditions of stress.

Pyridoxine (Vitamin B_6)

This plays a vital role in the balance of sodium and potassium in the body and so helps regulate body fluids. It protects against stress and nervous disorders being necessary to enzyme functions in that field. The requirement of this vitamin is increased by the use of oral contraceptives and other drugs.

Vitamin B_{12}

It is necessary for normal metabolism, for nerve tissue and for some forms of anaemia, and also plays an important role in proper digestion, assimilation and elimination.

Biotin

This helps to build resistance to allergies, and although it can be manufactured in our intestines it is destroyed by antibiotics.

Choline

This is essential for the health of the liver and kidneys.

Inositol

This protects the liver, kidneys and heart and is needed for proper growth of bone marrow and is essential for hair growth.

Calcium Pantothenate

This stimulates the adrenal gland, is important for healthy nerves and skin, helps to withstand stress and is helpful with premature ageing processes such as combating wrinkles.

Vitamin C

This is a protective vitamin essential to overall health. It aids in resisting virus and bacterial infection and helps to promote healthy teeth and gums. It also helps to reduce the effects of some allergies.

Vitamin E

This promotes the function of the sexual glands and activity of the reproductive system. It improves circulation, increases the production of necessary hormones, and revitalizes body cells.

Calcium

The main function of this mineral is to work with the phosphates to build and maintain nerves, bones, teeth and to help resistance to infection, to promote healing and it aids the utilization of iron.

Copper

This mineral is essential in promoting blood cell production and increases tissue respiration.

Iodine

This is essential in the development and functioning of the thyroid gland, helps resist infections and relieves nervous tensions.

Iron

This is the essential mineral for the blood to attract and transmit oxygen throughout the body, so that it enriches the blood and feeds the tissues.

Magnesium

This mineral calms the nerves, maintains the

elasticity of the tissues and helps promote absorption of other minerals such as calcium, phosphorus, sodium and potassium.

Manganese

This helps nourish the nerves and brain, promote milk production in pregnancy, and helps maintain sex hormone production.

Phosphates

These are essential for building proper brain and nervous tissue and maintaining these in optimum condition. Also promotes secretion of necessary hormones.

Potassium

This mineral works together with sodium in controlling the body fluids. It helps in disorders of the lungs and chest, and also with liver disorders.

Zinc

This mineral aids digestion and assists in metabolism of other minerals. It is helpful with hormone production and essential for a healthy prostate gland.

* * *

So, to summarize, the *New Era* range of vitamin preparations contain properly prepared tissue salts just as those in the biochemic remedies. The multi-vitamin preparation is an all purpose supplement providing the important vitamins and minerals in easy to swallow capsule form. They are combined with *Ferr. Phos.*, *Kali. Phos.* and *Mag. Phos.*

The Vitamin B Complex combines the important B vitamins combined with the tissue salts *Calc. Phos.*, *Ferr. Phos.*, *Mag. Phos.* and *Kali. Phos.*

The Vitamin C Capsules combine this essential vitamin, which the body cannot store, together with the tissue salts *Calc. Phos.*, *Ferr. Phos.*, *Mag. Phos.* and *Kali. Phos.*

The Vitamin E Capsules contain the vital hormone combined with the *New Era* tissue salts *Calc. Phos.*, *Ferr. Phos.*, *Mag. Phos.*, *Kali. Phos.*

See the Therapeutic Index for some of their uses in common ailments.

6.

MINERALS AND TRACE ELEMENTS

Dr Schuessler was a pioneer in the realization of the importance of supplying the body with the right minerals in the right form to maintain good health and to correct ill-health.

Now nearly everybody recognizes the importance to health, growth and life of Calcium and Iron. It has also been discovered that some elements are dangerous in excess, especially Sodium in the form of common salt. We have begun to understand the importance of the so-called trace elements and it is generally realized that for complete health the body requires a supply of practically every inorganic element, mostly in very small quantities.

During the 50's and 60's the public became gradually aware of the importance of many vitamins to health, and it seems a safe prediction that throughout the 80's and 90's medical science will progress greatly in its knowledge of the function and value of trace elements, and this knowledge will be made known to the general public.

Meanwhile, there is enough information to give in this chapter a broad outline of the importance of the various minerals and trace elements drawing particular attention to those where deficiencies may arise. With this knowledge readers can consider for themselves whether they should take action to guard themselves against deficiencies by modifying their diet or by taking supplements, or both.

Calcium and Magnesium

The first two major minerals are Calcium and Magnesium. These are mentioned together because they work together in various ways, especially in maintaining the health of our bones and our muscles. They are very important indeed in maintaining the muscles of the heart in good condition.

Those of us who live in hard-water districts and drink tap-water probably get a good supply of both of these elements from that source. Dairy products such as milk, cheese and yogurt are a very valuable source of Calcium. Butter and cream do not provide very much as most of the Calcium is left behind in the skimmed milk. Skimmed milk powder, which can be used in cooking, in making yogurt, re-constituted as liquid milk etc., is a very valuable source of Calcium free of butter fat.

Vegetables are a good source of Magnesium especially if eaten raw. As a result vegetarians are unlikely to be short of Magnesium, but conversely vegans may be short of Calcium.

Fortunately, a very logical natural source of both Calcium and Magnesium is available in the form of Dolomite tablets. Dolomite is a mineral whose origin, like chalk, was in the laying down over millions of years of bones, shells, etc., of primitive creatures, and it has the important advantage that Calcium and Magnesium are present in it in the ratio of 1.65:1 which is exactly the same ratio as these elements are required in the body.

Iron

The next most important mineral to consider is Iron which, as everybody knows, is a vital constituent of blood, without which the blood cannot perform its essential function of transporting oxygen around the body. Lack of Iron eventually leads to anaemia.

Those of us eating a normal mixed diet including meats, and especially liver, probably get enough Iron. Vegetarians are at risk, especially vegans, unless they watch their diets very carefully making sure that they eat plenty of soya beans, dried fruits etc. Growing children and women of childbearing age have a larger requirement for Iron than the rest of us and are hence in danger of becoming deficient even on a normal mixed diet.

Iron supplements have been available for many years but unfortunately most of the earlier ones on the market were in forms which were poorly absorbed and could easily give rise to gastric irritation. Now, however, much improved Iron supplement tablets are available which provide this element in a form which is much more readily absorbed by the body and is less likely to cause gastric upsets. The best of these supplements are probably those where the Iron is present as gluconate.

Potassium and Sodium

The next two elements to be considered are Potassium and Sodium. Not only are both these elements required for the healthy functioning of the body but it is vital that we have them in the correct balance as it has been demonstrated that some people will develop high blood-pressure and other heart diseases if they have an excess of Sodium over a period of years.

We get most of our Sodium as common salt added to our food at the table and during the cooking process. Unfortunately, manufacturers also add Sodium, usually as common salt, to many processed foods during the course of manufacture. This is obvious in the case of products such as ham, bacon, etc., but perhaps not quite so obvious in very many other foodstuffs such as sausages, canned soups, butter, margarine, bread, etc.

The result of this is that anyone eating a normal Western diet will almost certainly be getting an excess of Sodium, probably a very large excess, unless they are making very strenuous efforts indeed to cut down on their salt intake. There is a great deal of very clear evidence of a direct link between the excess of Sodium in our diet and heart disease. For instance, the Belgians are very fond of ham, bacon and sausages of all kinds, and consume them in large quantities which results in a very high sodium intake. Over a period of time, however, they have developed more healthy eating habits and reduced their consumption of these salt-rich items. Happily, along with this reduction in salt intake, has come a reduced incidence of heart disease in Belgium.

We require rather more Potassium than Sodium in our diet, 2.5g per day as opposed to 2g per day of Sodium, and vegetables and fruits grown on soils containing a normal amount of Potassium are a good source. Unfortunately, food processing and cooking removes Potassium from our food with the result that it is estimated that only just enough Potassium for our needs is available in the average daily diet.

This is bad news for two reasons. One, if the average amount available is only just sufficient inevitably some of us are not getting enough. Also as we have seen the balance between Sodium and Potassium is important and as most of us are getting too much Sodium in our daily diet and only just enough or too little Potassium a very great number of us have a potentially dangerous ratio of Sodium to Potassium in our diet.

The first symptoms of a Potassium deficiency are a lack of energy, strength and drive generally. Unfortunately, an excess of Sodium both absolutely and in relation to Potassium does not show itself at once but possibly after some years in the form of high blood-pressure and heart disease.

What can we do about this? Well, first of all, all of us should be doing all we can to reduce the amount of

odium in our diet. It is so omnipresent in our food
s common salt that it is very unlikely that we will
ucceed in overdoing this unless we are living
ermanently or temporarily in a very hot climate or
ngaged in an occupation such as a steel
oundryman, or doing very strenuous sports, or in
ny other situation when we may lose a great deal of
alt in the form of perspiration. In these
omparatively rare situations we may actually have
o supplement our salt intake. In any case, we have
ur inbuilt physiological craving for salt which will
elp us to avoid the unlikely risk of our becoming
eficient in Sodium.

Potassium is another matter. As we have already
een, unless we are eating plenty of raw fruits and
egetables the chances are that we are barely getting
nough Potassium or are actually deficient in this
lement. Two questions arise, first is it dangerous to
ave an excess of Potassium, and second, how can
ve set about adding Potassium to our diet? To
nswer the first, there is probably no danger in a
noderate excess of Potassium in the diet, but if we
ecide to take a Potassium supplement it is
mportant which one we choose — more on this later.

There is even some evidence which suggests that
s most of us are getting far too much Sodium and
re thus to a certain extent at risk from the point of
iew of high blood-pressure and heart disease, a
noderate excess of Potassium may well have the
eneficial effect of countering this.

This can be inferred from the results of a recent
tudy in which young healthy males on normal diets

were given Potassium supplements in quantities equivalent to their normal total daily requirement. In other words if their diet was already providing them with the required amount of Potassium, their total intake following the addition of the Potassium supplement was double their normal requirement. In this admittedly short test no adverse effects were seen but significant reductions in blood-pressure were achieved.

How do we go about getting sufficient Potassium? Well, farmers put Potassium on their fields in the form of manufactured fertilizers, so most raw vegetables and fruits are a good source of Potassium. Bananas and oranges are a good source but if you looked to these fruits for your total requirement of Potassium you would have to eat ten Bananas or ten oranges a day to make sure of your total requirements. This will probably be too much of these particular fruits for most people.

Potassium supplements are available on the market but should be treated with caution. One type of tablet is made so as to release the Potassium slowly into the digestive system but unfortunately this type of tablet produces gastric irritation which can be dangerous. The other type is an effervescent tablet which is dissolved in water which is then drunk. This type of supplement avoids the problem of gastric irritation and makes the Potassium immediately available to the body instead of slowly over a period of time as would be the case when digesting a Potassium-rich food. The effervescent type of supplement is probably safe for most people to take providing no more than one tablet per day is taken immediately before or after a meal. Anyone in doubt, and this includes anyone with blood-pressure, heart disease or kidney problems, should consult a doctor before taking a Potassium supplement.

Trace Elements

We have so far considered five important metallic elements which are required by the body in quite significant quantities. We now come on to the question of trace elements which are required by the body in minute amounts, but nevertheless essential for complete health. The list of these trace elements is surprisingly long and includes virtually all the inorganic elements.

There are some surprises in this list! Arsenic, long known as a deadly poison, is in fact essential to the body in very minute quantities. Fortunately, there is Arsenic in our food and water in very minute amounts and a person with an Arsenic deficiency would be by way of being a curiosity. Thus we do not have to face up to the rather frightening prospect of trying to add more Arsenic to our diet!

The list of the trace elements which we require is a long one and includes Cobalt, Fluorine, Iodine, Nickel, Tin, Vanadium and quite a host of others. All are vital, but fortunately most of them we can get in one way or another in sufficient quantities from what we eat and drink. Tea for instance is a good source of fluorine and liver, kidneys and fish are good sources of most trace elements.

There are, however, five vital elements where

there is definite evidence of wide-spread deficiencies
These are as follows:

1. *Chromium.* This is essential for the maintenanc
of correct blood sugar, and a deficiency may b
associated with certain types of heart disease.
2. *Copper.* This is an essential part of variou
enzyme systems and a deficiency may wel
contribute to anaemia but an excess can b
dangerous.
3. *Selenium.* This is beginning to be regarded as
playing a valuable role in protection against
cancer, and has recently shown to be of value t
sufferers from arthritis.
4. *Zinc.* Probably the most important of all. It play
an essential part in more than forty of the body'
vital enzyme systems and is crucial for health
growth. Symptoms of a deficiency includ
depression, loss of appetite, poor or reduced se
drive and loss of taste and/or smell.
5. *Manganese.* This element is necessary for man
vital processes including liver metabolism, th
control of nervous irritability and the utilizatio
of Vitamin C.

We tend to become deficient in these four element
because our food is now grown on soils that b
successive cropping have been depleted of thes
minerals and it is not a normal farming practice t
put them back in the form of fertilizers. There ar
further losses in the cooking and processing of food
and vegetarians, particularly vegans, are especiall

t risk because the two richest sources of virtually all he trace elements are liver and fish.

Fortunately, supplements including Zinc and Manganese as gluconates and Selenium and Chromium as yeasts are available as a combined ablet with or without Iron. Also for those identified as deficient in copper a tablet containing Zinc and Copper gluconate in an optimum ratio is available.

I will sum up this chapter by trying to answer two questions. How do I go about making sure that I and ny family do not suffer from mineral or trace element deficiency? And how do I know if I have succeeded?

To answer the first, there are a few simple rules which help greatly:

1. If you are fortunate enough to live in a hard-water area, (I say fortunate because it has been demonstrated very clearly that mortality from heart disease is much lower in hard-water areas than soft), make sure that you and your family drink tap-water and use tap-water for cooking and making beverages such as tea or coffee. If you have a water softener it is fine to use softened water for bathing, washing clothes etc., but do not drink it. The valuable Calcium and Magnesium have been removed and sinister Sodium introduced in their place.
2. Unless you do live in a very hot climate, engage in very considerable physical exertion or otherwise are one of the very few where Sodium depletion through excessive perspiration is a

danger, do your very best to reduce the amoun
of salt you take at table, use in cooking or take i
the form of processed foods.

3. Eat a mixed diet containing as little as possibl
animal fat but with an emphasis on natura
unprocessed foods, uncooked fruits an
vegetables and cereal products. Non-fat dair
products such as skimmed milk and cottag
cheese and liver and fish are also very valuabl
sources of minerals and trace elements. If yo
are a vegetarian or vegan I respect you
convictions but would point out to you that yo
do have to watch your diet very carefully or tak
supplements to avoid deficiencies of Iron, an
Calcium and the trace elements.

4. Finally, comes the matter of supplements:
Dolomite tablets. These are a very logical, natura
and safe source of both Calcium and Magnesiun
— and you need have no hesitation in takin
them in the recommended dosage if you thin
you may have a deficiency of either Calcium o
Magnesium.
Potassium. Unfortunately, many of us ar
deficient in Potassium and/or have an advers
Sodium/Potassium balance. If you suspect this i
the case you can eat oranges and bananas or tak
a Potassium supplement. Provided the correc
supplement is chosen, i.e. the effervescent table
which is drunk in the dissolved state. The takin
of a Potassium supplement when required b
persons in normal health and strictly followin
the dosage directions should present n
problems.

Iron. Safe and effective Iron supplements are now available and anyone particularly at risk of an iron deficiency especially women of childbearing age or strict vegetarians should have no hesitation in taking an iron supplement if they feel they need it.

Trace elements. As we have seen, we need a large number of trace elements in our diet but fortunately, one way or another, we get an adequate supply of most of these. It does seem, however, that many people may tend to be deficient of one or other of the elements Zinc, Manganese, Chromium, Copper and Selenium. Fortunately, these are now available as tablets in an easily assimilable form in one tablet, with or without Iron.

Finally, how do you know whether you and your family are getting all the minerals and trace elements that they should? If your normal diet is more or less as has been described above, and particularly if you are taking the supplements recommended where any doubt exists, all is probably well. Your best confirmation of this will come if everyone in your family is full of health and vitality and resistant to minor infections.

If, however, all is not well and you still have doubts there is now a valuable check in the form of hair analysis. It has long been known that the minerals circulating in our blood stream are laid down in our hair and finger nails. This fact has been used in forensic medicine to detect the presence of

Arsenic in the hair. Over the past few years the technique of hair analysis has been greatly improved as a means of seeing whether we have mineral deficiency especially of trace elements.

If you wish to take advantage of this service you may be interested to know that the first totally British facility has recently been established in London by New Era Laboratories Ltd. at 39 Wales Farm Road, North Acton, utilizing the experience and guidance of Dr P. Barlow of Aston University, Birmingham. This is no more expensive than other systems on offer which have to go to America for analysis, is perhaps even more advanced than some American systems and is much quicker in getting the results back to you. *New Era* will analyze your hair sample for you using a very modern computerized system and will send you a report which will show you how the mineral and trace element content of your hair compares with normal, and also warn you if the levels of any toxic metals such as lead or mercury are too high.

If this indicates any marked problem you can take corrective action either by modifying your eating habits or by taking appropriate supplements. Unfortunately, this service is not available on the N.H.S. Details are, however, available in most Health Food Stores.

7.

THE HYMOSA RANGE OF BEAUTY PREPARATIONS

These are an exclusive range of beauty preparations with biochemic tissue salts incorporated in them. They can be used with advantage by all women for the overall care of their skin, and those who work hard and have little time to spare will nevertheless find their time well spent from the results of using these preparations.

But the world of beauty has come to mean a lot more than superficial appearance. In order to look and be one's best, it must be realized that beauty and health go hand in hand. Good health shows itself in a clear skin, glossy hair, and a fit body. Being beauty conscious involves a degree of self-esteem, and most doctors feel that to try and be at one's best is a sign of a healthy, well-balanced person. We know from hospital experience that a woman's renewed interest in her appearance is an important sign of recovery from illness.

Our bodies reflect the ups and downs of diet, illness and stress, and our bodies changing patterns of health are reflected in the skin. But few women have a perfect skin and to achieve a good complexion will need care and patience. And for those women who fortunately do have a beautiful skin they must care for and protect it to safeguard its qualities.

The quality of our skin depends not only on how we care for its surface over the years but also on the health and condition of the skin's deeper layers. This naturally involves the proper intake of properly balanced healthy foods and the avoidance of excessive weight. The *New Era* biochemic tissue salt remedies play an important part in maintaining health, and the benefit from these can be reflected in the skin.

And, for the surface of the skin, the *Hymosa* range of preparations containing these vital tissue salts fits perfectly into the pattern of full over-all care. First cleansing of the skin is a top priority. Not to just remove what has previously been applied locally but also to remove general grime and to thoroughly open the pores for elimination. *Hymosa* soap is an excellent first stage cleanser for all the body, and subsequently *Hymosa Cleansing Milk* can, with benefit, be routinely used upon the face for thorough cleansing. Remember, this contains biochemic tissue salts and is suitable for all skin types and it leaves the skin naturally soft and supple.

The cleanser is usefully followed by the *Hymosa Skin Freshener Lotion* which contains extract of herbs which suitably improves the skin texture. Also the

Hymosa Lotion containing the vital tissue salts, used as a hand and body lotion, will keep the skin of the body soft, smooth and velvety.

Moisture is the vital balance in the upper layers of the skin and the *Hymosa Moisturizing Lotion*, containing biochemic tissue salts, moisturizes the skin and keeps it soft and smooth subsequent to the cleansing.

For night care the *Hymosa E Cream*, also containing biochemic tissue salts, is suitable for all ages. But also regular use of this night treatment vitamin E cream will help slow down the ravages of time.

For various skin problems, including the nails and hair, please see the Therapeutic Index.

Men are not forgotten. *Hymosa* soap is an excellent cleanser for all skins and the *Hymosa After Shave Cream* also contains the vital biochemic tissue salts to help the skin to optimum condition.

All of these products are unique in that they contain biochemic tissue salts which can only do good to the skin. So that regular use as directed should be of great benefit to all.

8.

THERAPEUTIC INDEX

Always continue to take your doctor's prescribed medicine in the usual way plus the tissue salts as appropriate. In many conditions the tissue salts alone can be found most helpful, but remember always see your doctor as necessary. The following alphabetical list of ailments gives the Combination Remedies advised and this is followed by the tissue salts which are also appropriate, given by number and by their Latin names.

Abdomen, bloated — Combination E; Combination S; No.7 (*Kali. Sulph.*); No.8 (*Mag. Phos.*).

Aching feet and legs — *Elasto*; Combination C; No.1 (*Calc. Fluor.*).

Aches and pains generally — Combination G; Combination A; Combination M.

Acid regurgitation — Combination S; No.10 (*Nat. Phos.*).

Acidity — Combination C; No.8 (*Mag. Phos.*).

Acne — Combination D; No.3 (*Calc. Sulph.*); *New Era* multi-vitamin capsules; and *Hymosa Balm*.

Ageing — Multi-vitamin capsules; Vitamin E capsules.

Anaemia — Combination B; multi-vitamin tablets; No.4 (*Ferr. Phos.*); No.2 (*Calc. Phos.*).

Anus, itching of, and cracks and fissures — *Elasto*; No.1 (*Calc. Fluor.*); No.3 (*Calc. Sulph.*).

Anxiety — *Nervone*; Vitamin B capsules; No.6 (*Kali. Phos.*).

Arthritis — *Zief*; Combination M; Combination I.

Asthma — Combination J; Combination H; No.6 (*Kali. Phos.*); No.8 (*Mag. Phos.*).

Appetite, abnormal — *Nervone*; No.6 (*Kali. Phos.*).

Appetite, poor — *New Era* multi-vitamin capsules.

Backache — *Zief; Elasto NR;* Combinations G; M; A; No.1 (*Calc. Fluor.*).

Beard area, pimples — Combination D; No.3 (*Calc. Sulph.*); *Hymosa Balm.*

Bed wetting in children — No.6 (*Kali. Phos.*); No.4 (*Fer. Phos.*); Combination B.

Belching — Combination C; Combination S; No.10 (*Nat. Phos.*).

Billiousness — Combination S; No.5 (*Kali. Mur.*).

Blisters — Combination K; No.9 (*Nat. Mur.*).

Blockage of tear ducts from cold weather — Combination Q ; No.9 (*Nat. Mur.*).

Bloodshot Eyes — Combination H; No. 4 (*Ferr. Phos.*).

Boils — Multi-vitamin capsules; Combination D; No.3 (*Calc. Sulph.*).

Bowels — See under constipation and diarrhoea.

Brittle nails — Combination K; No.7 (*Kali. Sulph.*).

Bronchitis — Vitamin C capsules; Combination J; No.4 (*Ferr. Phos.*).

Burns — Combination D; Combination M; Vitamin C capsules; No.5 (*Kali. Mur.*); No.3 (*Calc. Sulph.*)

Breath, offensive — Combination C; No.10 (*Nat. Phos.*).

Callossities — *Elasto*; *Elasto Balm*; Combination L; No.1 (*Calc. Fluor.*).

Catarrh — Combination Q ; No.7 (*Kali. Sulph.*).

Catarrh in ear, causing mild deafness — Combination Q ; No.7 (*Kali. Sulph.*); No.5 (*Kali. Mur.*).

Chapped hands from cold — Combination P; *Hymosa Balm*; No.1 (*Calc. Fluor.*).

Chapping of lips — Combination P; *Hymosa Balm;* No.1 (*Calc. Fluor.*); No.3 (*Calc. Sulph.*).

Chestiness — Combination J; No.5 (*Kali. Mur.*).

Chilblains on hands and feet — *Elasto Balm*; Multivitamin capsules; Combination P; No.6 (*Kali. Phos.*); No.2 (*Calc. Phos.*).

Chills and chilliness — Combination B; No.4 (*Ferr. Phos.*).

Circulation, poor — Combination N; No.6 (*Kali. Phos.*); No.2 (*Calc. Phos.*).

Colds — Combination J; No.4 (*Ferr. Phos.*); Vitamin C capsules.

Colic — Combination E; No.8 (*Mag. Phos.*).

Conditioner — Combination S; Combination D.

Constipation — Combination S; No.5 (*Kali. Mur.*).

Coughs — Combination J; No.4 (*Ferr. Phos.*);
acute and irritating — Combination J; No.4 (*Ferr. Phos*).
convulsive bouts — Combination H; No.8 (*Mag. Phos.*).

Cracking of joints — *Zief*; Combination M; No.10 (*Nat. Phos.*).

Cracks in palms of hand — Combination L; *Elasto* Balm; No.1 (*Calc. Fluor.*).

Cramp — *Elasto*; Multi-vitamin tablets; No.8 (*Mag. Phos.*).

Craving for food — *Nervone*; No.6 (*Kali. Phos.*).

Craving for salt or salty food — Combination Q ; No.9 (*Nat. Mur.*).

Cries easily — *Nervone*; No.6 (*Kali. Phos.*).

Cystitis — No.6 (*Kali. Phos.*); No.5 (*Kali. Mur.*); No.8 (*Mag. Phos.*); *Nervone*; Combination N.

Dandruff — Combination D; Combination K; No.7 (*Kali. Sulph*); No.9 (*Nat. Mur.*).

Depression and despondency — *Nervone*; No.6 (*Kali. Phos.*); Vitamin B capsules.

Diarrhoea — Combination S; No.5 (*Kali. Mur.*); No.10 (*Nat. Phos.*).

Digestive upsets — Combination C; No.11 (*Nat. Sulph.*).

Drooping of eyelids — *Elasto*; No.8 (*Mag. Phos.*); No.6 (*Kali. Phos.*).

Drowsiness — Combination F; No.12 (*Silica*).

Dryness of nose — Combination K; No.9 (*Nat. Mur.*); No.12 (*Silica*).

Dry skin — *Elasto*; *Elasto* Balm; Vitamin E capsules; No.1 (*Calc. Fluor.*).

Dysmenorrhoea — Combination N; No.8 (*Mag. Phos.*).

Dyspepsia — Combination C; No.10 (*Nat. Phos.*).

Earache — No.1 (*Ferr. Phos.*); No.7 (*Kali. Sulph.*); Combination J; Combination A.

Edginess — Combination B; No.6 (*Kali. Phos.*).

Enamel of teeth deficient — Combination R; No.1 (*Calc. Fluor.*); Multi-vitamins capsules.

Eneurisis — *See* bed-wetting.

Eyes water — Combination Q ; Combination H; No.9 (*Nat. Mur.*).

Eczema — Combination D; Vitamin C capsules.

Face, coldness — *Elasto*; No.2 (*Calc. Phos.*).
neuralgia with tears — Combination J; Combination A; Combination H; No.9 (*Nat. Mur.*).
neuralgia with shooting or spasmodic pains aggravated by cold — Combination A; No.8 (*Mag. Phos.*).

Faeces, difficult to expel — *Elasto*; No.1 (*Calc. Fluor.*).

Fainting, tendency to — *Nervone*, No.6 (*Kali. Phos.*).

Fatty food disagrees — Combination S; No.10 (*Nat. Phos.*); No.5 (*Nat. Mur.*).

Fever, and fever with chills and cramps — Combination I; No.4 (*Ferr. Phos.*); No.8 (*Mag. Phos.*).

Fibrositis — Combination I; *Zief*, No.4 (*Ferr. Phos.*).

Flatulence — Combination E; No.10 (*Nat. Phos.*).

Flu — Combination J; No.10 (*Nat. Sulph.*); No.4 (*Ferr. Phos.*); No.5 (*Kali. Mur.*); Vitamin C capsules.

Frequent passing of urine — *Nervone*; No.6 (*Kali. Phos.*).

Frigidity — *Nervone*, No.6 (*Kali. Phos.*); No.8 (*Mag. Phos.*); Vitamin E capsules.

Gastric indigestion — Combination C; No.10 (*Nat. Phos.*); Vitamin B capsules.

Gastric symptoms and billiousness — Combination E; No.8 (*Mag. Phos.*).

Gout — *Zief*; Combination A; Combination G; No.10 (*Nat. Phos.*); No.11 (*Nat. Sulph.*); No.4 (*Ferr. Phos.*).

Gum boil — Combination C; No.12 (*Silica*).

Gums, hot, swollen and inflamed — Combination Q; No.4 (*Ferr. Phos.*); No.3 (*Calc. Sulph.*); No.5 (*Kali. Mur.*).

Hair loss — Combination D; No.7 (*Kali. Sulph.*); No.12 (*Silica*); Multi-vitamin capsules.

Halitosis (bad breath) — Combination C; No.10 (*Nat. Phos.*); Vitamin B capsules.

Hard to wake in morning — *Elasto*; No.2 (*Calc. Phos.*).

Hay fever — Combination H; No.8 (*Mag. Phos.*); No.9 (*Nat. Mur.*).

Headache, billious type — Combination S; No.11 (*Nat. Sulph.*)
with pains — Combination F; Combination I; No.8 (*Mag. Phos.*).
nervous — *Nervone*; Combination F; No.6 (*Kali. Phos.*).

Heals slowly — Combination E; No.12 (*Silica*).

Haemorrhoids — *Elasto*; No.1 (*Calc. Fluor.*).

Heartburn — Combination C; No.11 (*Nat. Sulph.*); No.10 (*Nat. Phos.*).

Herpetic eruptions on skin — Combination D; No.9 (*Nat. Mur.*); *Hymosa* Balm.

Hiccups — Combination E; No.8 (*Mag. Phos.*).

Hip joint troubles — *Zief*; Combination M; Combination I.

Hoarseness — *Elasto*; No.2 (*Calc. Phos.*).

Hopelessness — *Nervone*, Combination F; No.6 (*Kali. Phos.*); No.9 (*Nat Mur.*).

Hysteria — *Nervone*; No.6 (*Kali. Phos.*).

Hot flushes — *Elasto*; No.4 (*Ferr. Phos.*); Vitamin E capsules.

Hyperactivity — *Nervone*; No.6 (*Kali. Phos.*).

Impatience and nervousness — *Nervone*; No.6 (*Kali. Phos.*).

Impotence — *Nervone*; No.6 (*Kali. Phos.*); Vitamin E capsules.

Incontinence — Combination B; Combination L; No.9 (*Nat. Mur.*); No.6 (*Kali. Phos.*).

Indigestion — Combination E; No.8 (*Mag. Phos.*).
with unpleasant taste in the mouth — Combination C; No.10 (*Nat. Phos.*).

Infant teething pains — Combination R; No.2 (*Calc. Phos.*); No.4 (*Ferr. Phos.*).

Inflammation — Combination L; Combination I; No.4 (*Ferr. Phos.*).

Influenza — Combination J; No.4 (*Ferr. Phos.*); No.5 (*Kali. Mur.*); No.10 (*Nat. Sulph.*); Vitamin C capsules.

Irritability — *Nervone*; No.6 (*Kali. Phos.*); Vitamin B capsules.

Itching of skin — Combination P; No.6 (*Kali. Phos.*); No.2 (*Calc. Phos.*); No.3 (*Calc. Sulph.*); *Hymosa* Balm; Vitamin E capsules.

Laryngitis — Combination J; No.2 (*Calc. Phos.*); No.4 (*Ferr. Phos.*).

Lethargy — Combination L; Vitamin B capsules.

Leucorrhoea — *See* vaginal discharge, white.

Lips cracked — *Elasto*; *Elasto* Balm; No.1 (*Calc. Fluor.*); No.3 (*Calc. Sulph.*).

Liverishness — Combination E; No.11 (*Nat. Sulph.*); Vitamin B capsules.

Loose stools — Combination E; No.11 (*Nat. Sulph.*).

Lumbago — *Elasto* NR; Combination G; Combination A; No.2 (*Calc. Phos.*).

Memory, poor — *Nervone*; No.6 (*Kali. Phos.*); Vitamin B capsules.

Menopausal symptoms, hot flushes — *Elasto*; No. 4 (*Ferr. Phos.*); Vitamin E capsules.
irritability — Combination B; No.6 (*Kali. Phos.*); Vitamin C capsules.
feeling run-down — Combination B; No.2 (*Calc. Phos.*); Multi-vitamin capsules.

Menstrual pain — Combination N; No.8 (*Mag. Phos.*); No. 6 (*Kali. Phos.*); *Elasto* NR.

Migraine — Combination F; No.11 (*Nat. Sulph.*); No.6 (*Kali. Phos.*).

Minor skin complaints — Combination D; No.7 (*Kali. Sulph.*); *Hymosa* Balm; Vitamin C capsules.

Moody, anxious — *Nervone*; No.6 (*Kali. Phos.*); Vitamin B capsules.

Morning sickness — Combination S; No.10 (*Nat. Phos.*); Multi-vitamin capsules.

Muscular pain — Combination I; Combination M; *Elasto* NR; No.8 (*Mag. Phos.*).

Nails, brittle — Combination K; Combination L; No.12 (*Silica*); No.7 (*Kali. Sulph.*); Vitamin E capsules.

Nails, inflammation around — Combination L; No.4 (*Ferr. Phos.*); Multi-vitamin capsules.

Nasal catarrh — Combination Q ; Combination H; No.4 (*Ferr. Phos.*).

Nasal discharge, clear, watery — Combination J; No.9 (*Nat. Mur.*).
thick — Combination Q ; No.5 (*Kali. Mur.*).

Nausea — Combination S; No.10 (*Nat. Phos.*).

Neck muscles, stiff — *Zief*; Combination M; Combination G; Combination I; No.4 (*Ferr. Phos.*).

Nerves and nerviness — *Nervone*; No.6 (*Kali. Phos.*); Vitamin B capsules.

Nervous and migraine headache — Combination F; No.6 (*Kali. Phos.*); No.8 (*Mag. Phos.*).

Nettle rash — Combination L; No.1 (*Calc. Fluor.*); No.3 (*Calc. Sulph.*); *Hymosa* Balm.

Neuralgia — *Nervone*; No.2 (*Calc. Phos.*).
with periods — Combination N; Combination A; No.8 (*Mag. Phos.*).

Nightmares — *Nervone*; No.6 (*Kali. Phos.*).

Nose bleed — No.4 (*Ferr. Phos.*).

Palpitations, nervous — *Nervone*; Combination B; No.6 (*Kali. Phos.*).

Periods, heavy — *Elasto*; No.4 (*Ferr. Phos.*); Multivitamin capsules.
painful — Combination N; No.8 (*Mag. Phos.*).
scanty — Combination L; No.9 (*Nat. Mur.*); No.3 (*Calc. Sulph.*); Vitamin C capsules.

Perspiration, excessive — *Nervone*; No.6 (*Kali. Phos.*); No.2 (*Calc. Phos.*).
lack of, to promote — Combination D; No.7 (*Kali. Sulph.*).

Pharyngitis — Combination R; No.2 (*Calc. Phos.*).

Phlegm, thick — *Zief*; No.12 (*Silica*).
watery — Combination J; No.9 (*Nat. Mur.*).

Piles — *Elasto*; No.1 (*Calc. Fluor.*).

Pimples — Combination D; No.3 (*Calc. Sulph.*); *Hymosa* Balm; Vitamin C capsules.

Pregnancy — *Elasto*; No.2 (*Calc. Phos.*); Multivitamin capsules.

Psoriasis — Combination D; Vitamin C capsules.

Pruritis — *See* itching of skin.

Pustules on face — Combination D; No.5 (*Kali. Mur.*); No.12 (*Silica*); Multi-vitamin capsules; *Hymosa* Balm; No.3 (*Calc. Sulph.*).

Queasiness — Combination S; No.10 (*Nat. Phos.*).

Retention of stools — Combination L; No.9 (*Nat. Mur.*).

Rheumatic pains — *Zief*; Combination M; *Elasto* NR; No.4 (*Ferr. Phos.*).

Rheumatism — *Zief*; *Elasto* NR; Combination I; No.10 (*Nat. Phos.*); No.11 (*Nat. Sulph.*).

Rhinitis — *See* nasal discharge and colds.

Runny nose — Combination J; No.9 (*Nat. Mur.*).

Run Down — Combination L; Vitamin B capsules.

Saliva, excess of — Combination Q ; No.9 (*Nat. Mur.*).
salty taste in mouth — Combination L; No.9 (*Nat. Mur.*).

Scabs on skin — Combination E; No.3 (*Calc. Sulph.*); Multi-vitamin tablets; *Hymosa* Balm.

Scaly skin eruptions — Combination D; No.7 (*Kali. Sulph.*).

Sciatica — *Zief*; *Elasto* NR; Combination A; Combination M; No.8 (*Mag. Phos.*); No.4 (*Ferr. Phos.*).

Sedentary life-style — Combination L; No.1 (*Calc. Fluor.*); Multi-vitamin capsules.

Sensitiveness — *Nervone*: No. 6 (*Kali. Phos.*).

Sexual difficulties — *Nervone*; No.6 (*Kali. Phos.*); Vitamin E capsules.

Sick headache from gastric upset — Combination S; No.11 (*Nat. Sulph.*).

Sickness (vomiting) — Combination C; No.10 (*Nat. Phos.*).

Shingles — Combination D; No.5 (*Kali. Mur.*) No.9 (*Nat. Mur.*); Vitamin C capsules.
for the pain — Combination A; Combination I; No.8 (*Mag. Phos.*).

Shivering — *Elasto*; No.2 (*Calc. Phos.*).

Shyness — *Nervone*; No.6 (*Kali. Phos.*).

Sick headache — Combination S; No.10 (*Nat. Phos.*); No.11 (*Nat Sulph.*).

Sinus disorders — Combination Q ; No.4 (*Ferr. Phos.*); No.5 (*Kali. Mur.*).

Skin, dry — Combination K; No.7 (*Kali. Sulph.*); *Hymosa* Balm.
greasy — Combination C; No.10 (*Nat. Phos.*); No.3 (*Calc. Sulph.*).
minor ailments — Combination D; No.5 (*Kali. Mur.*); Vitamin C capsules; *Hymosa* Balm.
scaling — Combination D; No.7 (*Kali. Sulph.*); *Hymosa* Balm; Vitamin E capsules.
spotty — Combination D; *Hymosa* Beauty preparations.
wrinkled — *Elasto*; No.6 (*Kali. Phos.*); Vitamin E capsules; *Hymosa* preparations.

Sleepiness — Combination L; No.9 (*Nat. Mur.*).

Sleeplessness from nervous causes — *Nervone*; No.6 (*Kali. Phos.*).

Smell, loss of — Combination J; Combination H; No.9 (*Nat. Mur.*).

Sneezing — Combination Q ; No.12 (*Silica*); No.9 (*Nat. Mur.*).

Sore lips — Combination D; No.3 (*Calc. Sulph.*); *Hymosa* Balm.

Sore throat — Combination J; No.4 (*Ferr. Phos.*).

Sour taste in mouth — Combination C; No.10 (*Nat. Phos.*).

Spasmodic twitching of eyelids — *Nervone*; No.8 (*Mag. Phos.*).

Sprains and strains — Combination I; Combination P; No.4 (*Ferr. Phos.*).

Stings of insects — Combination Q ; No.9 (*Nat. Mur.*).

Stomach upsets — Combination S; No.11 (*Nat. Sulph.*); No.10 (*Nat. Phos.*).

Stomatitis — Combination C; No.10 (*Nat. Phos.*); Vitamin C capsules.

Stuffy cold — Combination J; No.7 (*Kali. Phos.*); Vitamin C capsules.

Stye — Combination D; No.12 (*Silica*); Multi-vitamin capsules.

Sweating at night — Combination L; No.9 (*Nat. Mur.*).

Taste, loss of — Combination Q ; No.9 (*Nat. Mur.*); Multi-vitamin tablets.

Taste in mouth, unpleasant — Combination C; No.11 (*Nat. Sulph.*); No.10 (*Nat. Phos.*).

Tears when going into the cold air — Combination J; Combination H; No.9 (*Nat. Mur.*).

Teeth sensitive to cold and to touch — *Elasto*; No.8 (*Mag. Phos.*); No.1 (*Calc. Fluor.*).

Teething with drooling — Combination L; No.9 (*Nat. Mur.*).

Teething — Combination R; No.2 (*Calc. Phos.*).

Teething retarded — Combination R; No.2 (*Calc. Phos.*).

Thirst — Combination Q ; No.9 (*Nat. Mur.*).

Throat, sore — Combination J; No.4 (*Ferr. Phos.*).
sore and dry — Combination J; No.9 (*Nat. Mur.*).
sore in singers and speakers — *Elasto*; No.4 (*Ferr. Phos.*).
tickling in — *Elasto*; No.1 (*Calc. Fluor.*).

Throat ulcers — Combination C; No.3 (*Calc. Sulph.*); No.12 (*Silica*).

Tiredness from overwork — *Nervone*; No.6 (*Kali. Phos.*); Multi-vitamin tablets.

Tongue, blisters on — Combination Q ; No.9 (*Nat. Mur.*).
coated — Combination C; No.10 (*Nat. Phos.*).
inflamed — Combination Q ; No.4 (*Ferr. Phos.*).

Tonsils, inflamed — Combination Q ; No.4 (*Ferr. Phos.*); Vitamin C capsules.

Tooth enamel, rough — Combination L; No.1 (*Calc. Fluor.*); Multi-vitamin capsules.

Toothache — *Nervone*; Combination R; Combination A; No.8 (*Mag. Phos.*).

Travel sickness — Combination F; Combination S; No.10 (*Nat. Phos.*); No.6 (*Kali. Phos.*).

Tremblings of part of body — *Nervone*; No.8 (*Mag. Phos.*); Multi-vitamin tablets.

Ulcers in mouth — Combination C; No.10 (*Nat. Phos.*); No.11 (*Nat. Sulph.*); Vitamin B capsules.
at corner of mouth — Combination D; No.12 (*Silica*); No.3 (*Calc. Sulph.*); Vitamin C capsules.

Upper respiratory tract inflamation — Combination J; No.4 (*Ferr. Phos.*).

Urine retention, spasmodic — *Elasto*; No.8 (*Mag. Phos.*).

Vaginal discharge, white — Combination D; Combination Q ; No.5 (*Kali. Mur.*).
coloured — Combination B; No.6 (*Kali. Phos.*).

Varicose veins and ulceration — *Elasto*; No.1 (*Calc. Fluor.*).

Vertigo — *Nervone*; Combination S; No.6 (*Kali. Phos.*); No.11 (*Nat. Sulph.*).

Voice, loss of, from strain — *Elasto*; No.4 (*Ferr. Phos.*).

Vomiting — Combination E; No.2 (*Calc. Phos.*).
sour, acid fluids — Combination C; No.10 (*Nat. Phos.*).

Warts — Combination D; Combination K; No.7 (*Kali. Sulph.*); No.5 (*Kali. Mur.*).

Water Brash — Combination L; No.9 (*Nat. Mur.*).

Weakness, convalescence — Combination B; No.2 (*Calc. Phos.*); Multi-vitamin capsules.

Weeps easily — *Nervone*; No.6 (*Kali. Phos.*); Vitamin B capsules.

Wheeziness — Combination Q ; No.5 (*Kali. Mur.*); No.4 (*Ferr. Phos.*); Vitamin C capsules.

Windpipe, spasms — *Elasto*; No.8 (*Mag. Phos.*).

REMEMBER IF YOU ARE EVER WORRIED ABOUT ANY SYMPTOMS ALWAYS CONSULT YOUR DOCTOR.

9.

CASE HISTORIES

Mr. T.S., aged 53, was made redundant two years ago, and after six months found further employment, but then that firm collapsed, so he is now out of work with no job prospects at all. This made him so stressful that he was put on tranquillizers by his doctor, the latter being a family friend and full of understanding.

However, the patient was fearful of his tranquillizers and had prescribed for him *Nervone* which made all the difference. He is now with some confidence taking steps to be self-employed and feels his stress has changed from hopelessness to taking action himself in a positive way, and he puts it down to the use of *Nervone*. In many cases stress and strains of life respond to these biochemic tissue salts.

Miss W.M., aged 27, had had spotty skin since adolescence, and was told she would grow out of it, but this had not happened so far. She found the skin condition most embarrassing. She had prescribed for her Combination D, plus suggested use of the range of *New Era Hymosa* Beauty/Skin preparations. Only a few weeks were needed to make a great difference. She is delighted and feels a new person.

* * *

Mrs G.M., aged 51, whose periods stopped at 48 years and had had flushes and menopausal symptoms distressing her since that time. Hormone tablets upset her and nothing else was offered. She had prescribed for her *Elasto,* No. 4 (*Ferr. Phos.*) and the vitamin E capsules. She responded in a few weeks and now flushes are rare, instead of being frequent as previously.

* * *

Miss D.M., aged 24, was always tired and lethargic and in a completely run-down condition. She was very efficient in the past but for the last two years has felt useless and has had no energy at all. She had prescribed for her Combination L, and vitamin B capsules and responded within days and has not looked back.

* * *

Mr D.S., aged 54, was always catching head colds, had prescribed for him Combination J, and vitamin C capsules. Within six months, he became totally clear and has not had a head cold for over a year whereas previously they were almost fortnightly.

* * *

Mrs A.V., aged 31, has had pre-menstrual tension since puberty and with hormone treatment and tablets she had no real results at all. She had prescribed for her *Nervone,* No. 6 (*Kali. Phos.*) and No. 9 (*Nat. Mur.*) with excellent results. She feels that by continuing with these tissue salts her symptoms are now minimal and no trouble.

* * *

Master D.M., aged 11, had had asthmatic attacks for several years which were getting worse each time. He was under the care of the hospital clinic and his excellent local doctor, but he had unfortunate side-effects from treatment. Although such treatment is essential for him, tissue salts were prescribed in addition. These were Combination J and Combination H, and his parents say that he seemed to improve markedly within a few weeks. These were continued and the tissue salts have meant that he is markedly improved, has an improved school attendance record. Similar help has been given to many adults also.

* * *

Mr P.H., aged 40, had had a spastic colon for many years. He had been fully investigated and treated by his doctor and by a first-rate hospital, but his condition had remained almost unchanged. He was prescribed Combination E, and No. 1 (*Calc. Fluor.*)

and within three months the condition had almost cleared.

* * *

Miss M.C., aged 19, had suffered painful periods ever since they began. Pain killers had too many side-effects to be tolerated in her case. She was prescribed Combination N for one week before the period was due, and she was helped with the first period after treatment. With continued use of this tissue salt in combination with Combination L her periods are now almost symptom free.

* * *

Dr D.V., aged 55, suffered from sciatica. He had had all the normal investigations by his colleagues in hospital and had been treated by experts in this field, but with no satisfactory result. He had prescribed for him Combination A only, but in doses of 5 tablets every four hours. He experienced relief within a day or so and repeated treatment keeps attacks in order. Now, in addition, he is taking *New Era* Vitamin E capsules to see if the ageing process behind the reasons for his case can be helped.

* * *

Mrs A.S., aged 34, wanted to become pregnant but had not conceived after ten years of marriage. Investigation of the couple indicated all was physically well with them both, but she was obviously under great stress from her inability to have a family. She took biochemic tissue salts No. 6 (*Kali. Phos.*) and *Nervone* to help her relax and No. 2

(*Calc. Phos.*) and *New Era* multi-vitamin capsules to help improve the general health of her body. She became pregnant after four months of treatment and continued with the No. 2 (*Calc. Phos.*) and the multi-vitamin tablets throughout the term, which helped to give her extra cover for her ante-natal care. The birth was excellent, the baby girl was breast fed, and now she is a toddler and her mother is happily pregnant again. During teething the baby was greatly helped by the *New Era* Combination R.

* * *

Mrs Z.F.A., aged 40, suffered migraine almost weekly, always on the left side, and she had had this for over twenty years and nothing seemed to help. She had had all the hospital investigations and usual medical treatment which unfortunately had not helped her at all. Therefore she came to biochemic tissue salt treatment and had prescribed for her Combination F, No. 6 (*Kali. Phos.*) which taken during an attack eased it to some extent. She continued taking Combination F and No. 6 (*Kali. Phos.*) over a period of one year and she was practically cured of her condition altogether. Instead of weekly attacks she has mild attacks, which she feels are of no importance, about once every two months. In a similar way relief was given to Mr E.C., aged 65, and Miss L.H., aged 23 years, and many others. Migraine can respond well to the biochemic tissue salts.

Mr J.W., aged 48, had had catarrh and sinus troubles for many years and nothing seemed to help despite full investigations by experts in this field. His stressful life aggravated the situation. The biochemic treatment given was Combination Q, and for his stress No. 4 (*Ferr. Phos.*), and No. 6 (*Kali. Phos.*). He gradually improved over a six month period and is now delighted. Treatment has been kept on and he has been almost clear since 1977.

* * *

Mr D.E., aged 31, suffered from hay fever and drugs proved of no avail, in fact they caused side-effects which made him even worse. He used *New Era* Combination H and also No. 9 (*Nat. Mur.*) at first for the wateriness. This soon helped him and cleared it completely. He takes the same regime of tablets each year before the hay fever season begins and throughout the season and is practically free from any attacks now.

* * *

Mrs A.J., aged 38, had difficulty sleeping as her mind was full of matters to do with her work as well as being overtired from running her home. She does not like sleeping pills because of fears of addiction, but has had to use them sometimes nevertheless. She had prescribed for her *Nervone,* and No. 6 (*Kali. Phos.*), and within days could relax at night, and her married life much improved. By continuing the tablets she now regularly sleeps well.

Miss H., aged 40, suffered from rheumatism in the small joints of her body which was getting very painful, and she feared the use of steroids. She had prescribed for her *Zief*, Combination M, Combination I, and *Elasto NR* when very painful. She progressed well and this encouraged her mother to try tablets for her arthritis. Her mother took these, but prescribed for herself, and she has also derived great benefit from them.

INDEX